Ten,
Of Course

Copyright © 2015 ORNA

Curl-Crested Jay couple
spotted two earthworms underneath their tree.
One is for Mrs. Jay, how many left for Mr. Jay ?

One, of course.

A couple are counted as two.

One worm for her and one for him,
together, how many ?

Two, (of course).

A couple of fish and their little fish have diamonds:
Two blue diamonds and one white.

How many diamonds do they have?

Three, of course.

One white diamond and two blue diamonds,
together they are:

Three, (of course).

There are beautiful roses in the garden.
Three butterflies and one sweet bee
are on their way to sip from their nectar.
There are just enough roses in the garden for them.

How many roses in the garden?

Four, of course.

One rose for the sweet bee
and three roses for the three butterflies,
together there are:

Four, (of course).

The Kindergarteners planted four Pomegranate trees
all around their Carob tree.

How many trees in their kindergarten now?

Five, of course.

One Carob tree and four Pomegranate trees,

together they are:

Five, (of course).

Five village kittens went out for a stroll in the city.
They met a friendly city cat,
who invited them to join him for an icy, tasty Pellegrino.
How many cups did they need?

Six, of course.

One for the city cat and five for the kittens,
together they were:

Six, (of course).

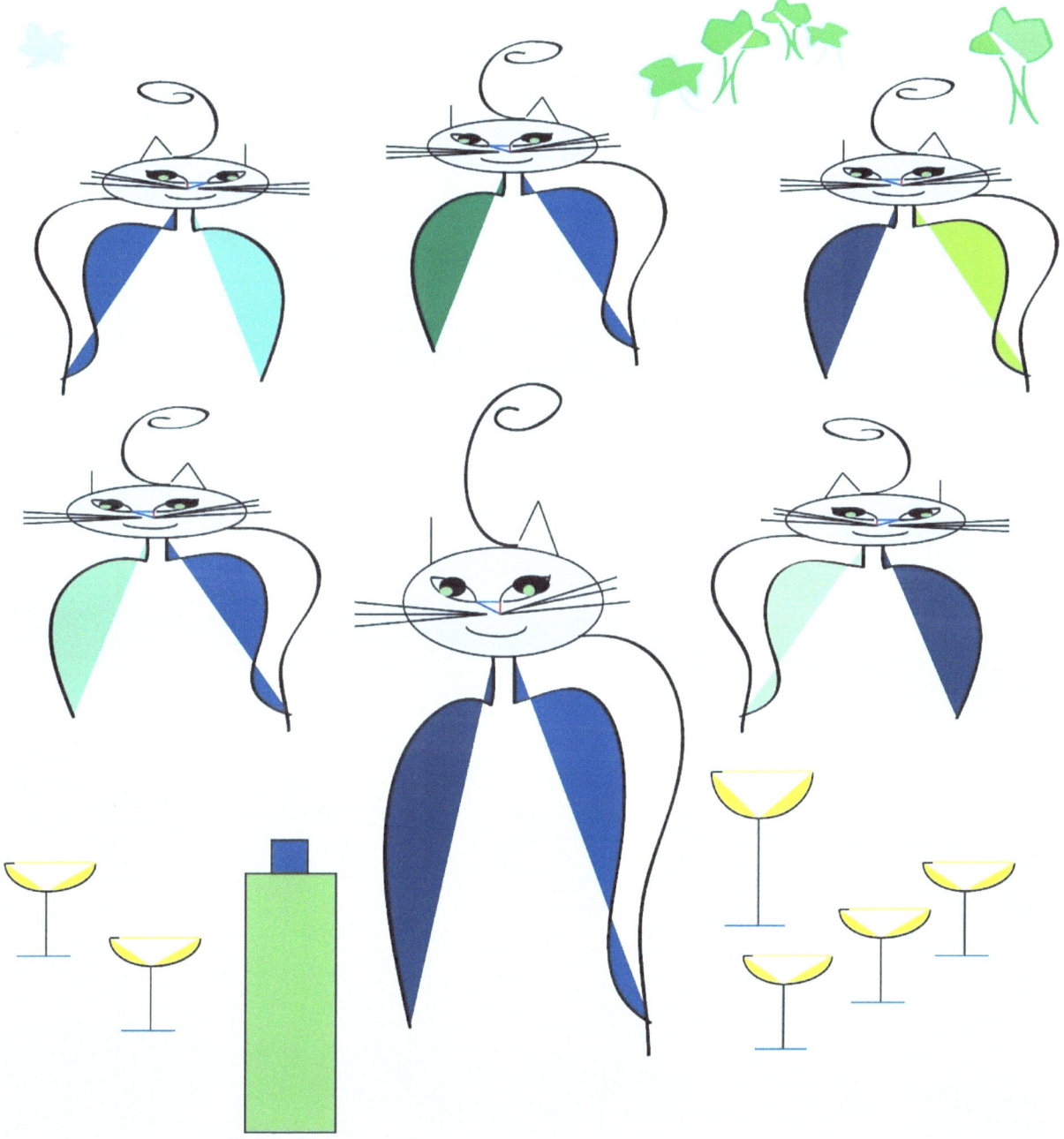

Before dinner, during sunset,
six turtles were walking for their pleasure.
Their friend, Mr. Rabbit, met them on their stroll
and invited them
to join him for fresh carrots from his garden.

How many carrots did they eat,
if each was hungry for only one sweet carrot ?

Seven, of course.

One carrot for Mr. Rabbit and six carrots for the six turtles,
together they were:

Seven, (of course).

The seven dwarves from Snow White story
picked one delicious apple each,
and together they chose
one most beautiful apple for Snow White.

How many apples did they have in their basket ?

Eight, of course.

One for Snow White and seven apples for themselves,

together they were:

Eight, (of course).

Eight ants are carrying a delicious cake with much care.
When they will arrive to their colony
they will need to share it with their queen.

To how many slices will they cut the cake?

Nine, of course.

One slice for the queen and eight between themselves.

together they are:

Nine, (of course).

Topaz is celebrating her Birthday and happiness is all around.
On the gorgeous cake are nine tall purple candles
and an additional one in green for next year.

How many candles on Topaz's cake?

Ten, of course.

One green candle and nine purple candles,
together they are:

Ten, (of course).

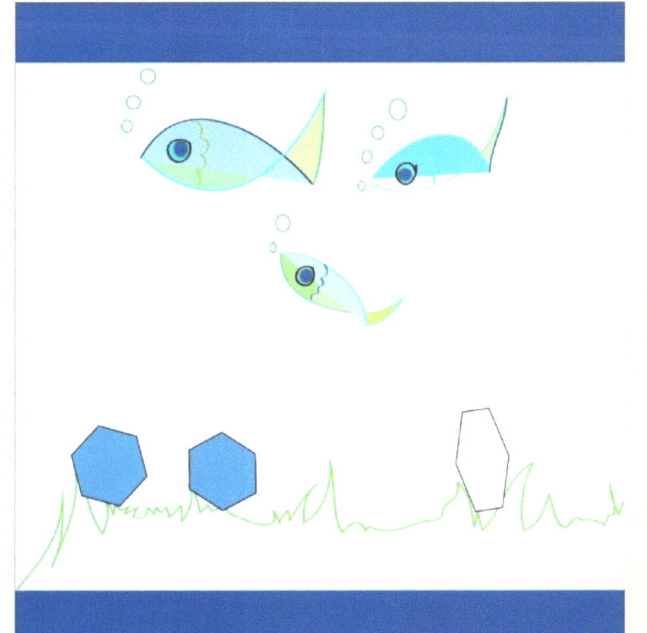

Written and illustrated by ORNA

www.ingramcontent.com/pod-product-compliance
Lightning Source LLC
Chambersburg PA
CBHW050432180526

45159CB00006B/2508